Jackson Hole
Total Eclipse Guide

Commemorative Official Guidebook

Aaron Linsdau

Sastrugi Press

Jackson Hole

Copyright © 2017 by Aaron Linsdau

All rights reserved. No part of this book may be reproduced or transmitted in any form or by any means, electronic or mechanical, including photocopying, recording, or by any computer system without the written permission of the author, except where permitted by law.

Sastrugi Press / Published by arrangement with the author

Jackson Hole Total Eclipse Guide: Commemorative Official Guidebook

The author has made every effort to accurately describe the locations contained in this work. Travel to some locations in this book is hazardous. The publisher has no control over and does not assume any responsibility for author or third-party websites or their content describing these locations, how to travel there, nor how to do it safely. Refer to official forest and national park regulations.

Any person exploring these locations is personally responsible for checking local conditions prior to departure. You are responsible for your own actions and decisions. The information contained in this work is based solely on the author's research at the time of publication and may not be accurate in the future. Neither the publisher nor the author assumes any liability for anyone climbing, exploring, visiting, or traveling to the locations described in this work. Climbing is dangerous by its nature. Any person engaging in mountain climbing is responsible for learning the proper techniques. The reader assumes all risks and accepts full responsibility for injuries, including death.
Park maps are courtesy of the National Park Service.

Sastrugi Press
PO Box 1297, Jackson, WY 83001, United States
www.sastrugipress.com
Quantity sales: Special discounts are available on quantity purchases by corporations, associations, and others. For details, contact the publisher at the address above.

Library of Congress Catalog-in-Publication Data
Library of Congress Control Number: 2017901278
Linsdau, Aaron
Jackson Hole Total Eclipse Guide / Aaron Linsdau- 1st United States edition
p. cm.
1. Nature 2. Astronomy 3. Travel 4. Photography
Summary: Learn everything you need to know about viewing, experiencing, and photographing the total eclipse in Jackson Hole on August 21, 2017.

ISBN-13: 978-1-944986-04-9
ISBN-10: 1-944986-04-9

508.4—dc23

Printed in the United States of America

All photography, maps and artwork by the author, except as noted.

10 9 8 7 6 5 4 3 2

Contents

Introduction	4
All About Eclipses	**6**
Total vs Partial Eclipse	7
Early Myth & Astronomy	9
Contemporary American Solar Phenomena	11
Future American Eclipses	13
All About Jackson	**14**
Overview of Jackson	14
Weather	19
Normal Jackson Weather Pattern	20
Forest Fires	21
Local News Information	22
Grand Teton National Park Safety	24
Altitude Sickness	25
Viewing and Photographing the Eclipse	**27**
Planning Ahead	28
Understanding Sun Position	29
Eclipse Data for Jackson, WY	31
Eclipse Photography	**31**
Eclipse Photography Gear	34
Camera Phones	36
Viewing Locations Around Jackson Hole	**43**
Notes	**64**
Remember the Jackson Hole Total Eclipse	**65**

Introduction

Thank you for purchasing this book. It has everything you need to know about the total eclipse in Jackson, Wyoming, on August 21, 2017.

A total eclipse passing across the United States is a rare event. The last U.S. total eclipse was in 1979. It traveled over Washington, Oregon, Montana, and the corner of North Dakota.

The next total eclipse over the U.S. will not be until April 8, 2024. It will pass over Texas, the Midwest, and on to Maine. After that, the next coast-to-coast total eclipse will be in 2045!

Now is the time. It's imperative to make travel plans today. You will be amazed at the number of people swarming to the total eclipse path. Some might say watching a partial versus a total eclipse is a similar experience. It's not.

This book is written for Jackson Hole visitors and anyone else viewing the eclipse. You will find general planning, viewing, and photography information inside. Should you travel to Jackson in mid-August, be prepared for an epic trip. Jackson Hole was beyond capacity in the summer of 2016. It had almost 5 million visitors over the summer months.

Foreign visitation has soared, too. The number of tour buses pouring into town routinely makes the news. This has become the new norm. Many Europeans take their vacation month in August and some will likely travel to see the total eclipse in America.

The Sturgis motorcycle rally will end one week before the eclipse. The event had a half million attendees in 2015. Bikers often drive to Jackson before and after the event. Chances are these bikers will travel to Wyoming.

All hotels in the Wyoming cities of Riverton, Casper, and Douglas are sold out as of the writing of this guide. Finding hotels or motels with available rooms along the eclipse path will be a major challenge.

Resources will be stretched far beyond the normal limits. Think gas lines from the late 1970s. It's common for the shelves of Jackson's four grocery stores to quickly become short on bread, bananas, and similar staples during the summer. Be prepared with backup supplies.

Jackson is a remote town and far from any major city. Wyoming roads are slow. The maximum speed limit is 55MPH near Jackson. Please obey posted speed limits within Grand Teton National Park. Jackson locals joke about people asking why it takes two hours to drive to Yellowstone National Park. "It's not that far," is the common refrain. It takes longer to drive anywhere in northwest Wyoming.

People in Jackson, Dubois, Driggs, Victor, and surrounding communities plan to rent out their properties for the eclipse. With a major celestial event in the summer of 2017, be assured that Jackson "hasn't seen anything yet."

Is this to say to avoid Jackson during the eclipse? Not at all! This guidebook provides ideas for interesting, alternative, and memorable locations to see the eclipse. It will be too late to rush to a better spot once the eclipse begins. The roads will likely be gridlocked. Law enforcement will be out to help drivers reconsider speeding.

Please be patient and watch out. Large, beautiful, but dangerous animals make the forests around Jackson their home. Throughout the year, over one hundred animals are hit in Grand Teton alone. Moose, bear, and bison are difficult to see along the highways.

You should feel compelled to take off work on August 21. Play hooky from your job. If your kids are in school, take them out. They'll likely be adults before the next chance to see a total eclipse. Create some family memories that will last a lifetime. The staff at Sastrugi Press normally does not advocate being irresponsible by skipping school or work. Make an exception because this is too big an event to miss.

Whether you visit Jackson or somewhere else along the total eclipse path, leave early and remember your eclipse glasses. People from all around the planet will converge on Jackson. Be good to your fellow humans and be safe. We all want to enjoy this spectacular show.

For the latest updates on this book, visit:

www.sastrugipress.com/jheclipse

Visit the following official websites for eclipse updates:

tetoneclipse.com

www.nps.gov/grte

www.fs.usda.gov/btnf

All About Eclipses

How an Eclipse Happens

An eclipse occurs when one celestial body falls in line with another, thus obscuring the sun from view. This occurs much more often than you'd think, considering how many bodies there are in the solar system. For instance, there are over 150 moons in the solar system. On Earth, we have two primary celestial bodies: the sun and the moon. The entire solar system is constantly in motion, with planets orbiting the sun and moons orbiting the planets. These celestial bodies often come into alignment. When these alignments cause the sun to be blocked, it is called an eclipse.

For an eclipse to occur, the sun, Earth, and moon must be in alignment. There are two types of eclipses: solar and lunar. A solar eclipse occurs when the moon obscures the sun. A lunar eclipse occurs when the moon passes through Earth's shadow. Solar eclipses are much more common, as we experience an average of 240 solar eclipses a century compared to an average of 150 lunar eclipses. Despite this, we are more likely to see a lunar eclipse than a solar eclipse. This is due to the visibility of each.

For a solar eclipse to be visible, you have to be in the moon's shadow. The problem with viewing a total eclipse is that the moon casts a small shadow over the world at any given time. You have to be in

* ILLUSTRATION NOT TO SCALE

a precise location to view a total eclipse. The issue that arises is that most of these locations are inaccessible to most people. Though many would like to see a total solar eclipse, most aren't about to set sail for the middle of the Pacific Ocean. In fact, a solar eclipse is visible in the same place on the world on average every 375 years. This means that if you miss a solar eclipse above your hometown, you're not going to see another one unless you travel or move.

It's much easier to catch a glimpse of a lunar eclipse, even though they occur at a much lower frequency than their solar counterparts. A lunar eclipse darkens the moon for a few hours. This is different than a new moon when it faces away from the sun. During these eclipses, the moon fades and becomes nearly invisible.

Another result of a lunar eclipse is a blood moon. Earth's atmosphere bends a small amount of sunlight onto the moon turning it orange-red. The blood moon is caused by the dawn or dusk light being refracted onto the moon during an eclipse.

Lunar eclipses are much easier to see. Even when the moon is in the shadow of Earth, it's still visible throughout the world because of how much smaller it is than Earth.

TOTAL VS. PARTIAL ECLIPSE

What is the difference between a partial and total eclipse? A total eclipse of either the sun or the moon will occur only when the sun, Earth, and the moon are aligned in a perfectly straight line. This ensures that either the sun or the moon is partially or completely obscured.

In contrast, a partial eclipse occurs when the alignment of the three celestial bodies is not in a perfectly straight line. These types of eclipses usually result in only a part of either the sun or the moon being obscured. This was often what led to ancient civilizations believing that some form of magical beast or deity was eating the sun or the moon. It appears as though something has taken a bite out of either the sun or the moon during a partial eclipse.

Total eclipses, rarer than partial eclipses, still occur quite often. It's more difficult for people to be in a position to experience such an event firsthand. Total solar eclipses can only be viewed from a small portion of the world that falls into the darkest part of the moon's shadow. Often this happens in the middle of the ocean.

The Moon's Shadow

The moon's shadow is divided into two parts: the umbra and the penumbra. The former is much smaller than the latter, as the umbra is the innermost and darkest part of the shadow. The umbra is thus the central point of the moon's shadow, meaning that it is extremely small in comparison to the entire shadow. For a total solar eclipse to be visible, you need to be directly beneath the umbra of the moon's shadow. This is because that is the only point at which the moon completely blocks the view of the sun.

In contrast, the penumbra is the region of the moon's shadow in which only a portion of the light cast by the sun is obscured. When

Total eclipse shadow 2016 as seen from 1 million miles on the Deep Space Climate Observatory satellite. Courtesy of NASA.

standing in the penumbra, you are viewing the eclipse at an angle. In the penumbra, the moon does not completely block the sun from view. This means that while the event is a total solar eclipse, you'll only see a partial eclipse. The umbra for the August 21 eclipse is approximately sixty miles wide. The penumbra will cover much of the United States.

To provide some context, the last total solar eclipse we experienced occurred on March 9, 2016, and was visible as a partial eclipse across most of the Pacific Ocean, parts of Asia, and Australia. However, the only place in the world to view this total solar eclipse was in a few parts of Indonesia.

Due to the varied locations and the brief periods for which they're visible, it's difficult to see each and every eclipse that occurs. Many people don't even realize that they have occurred. Consider that the umbra of the moon represents such a small fraction of the entire shadow and the majority of our planet is comprised of water. Thus, the rarity of being able to view a total solar eclipse increases significantly because it's likely that the umbra will fall over some part of the ocean rather than a populated landmass.

ECLIPSES THROUGHOUT HISTORY

Ancient peoples believed eclipses were from the wrath of angry gods, portents of doom and misfortune, or wars between celestial beings. Eclipses have played many roles in cultures, spawning myths since the dawn of time. Both solar and lunar eclipses affected societies worldwide. Inspiring fear, curiosity, and the creation of legends, eclipses have cast a long shadow in the collective unconscious of humanity throughout history.

EARLY MYTH & ASTRONOMY

Documented observations of solar eclipses have been found as far back in history as ancient Egyptian and Chinese records. Timekeeping was important to ancient Chinese cultures. Astronomical observations were an integral factor in the Chinese calendar. The first observation of a solar eclipse is found in Chinese records from over 4,000 years ago. Evidence suggests that ancient Egyptian observations may predate those archaic writings.

Many ancient societies, including Roman, Greek and Chinese civilizations, were able to infer and foresee solar eclipses from astronomical data. The sudden and unpredictable nature of solar eclipses had a stressful and intimidating effect on many societies that lacked the scientific insight to accurately predict astronomical events. Relying on the sun for their agricultural livelihood, those societies interpreted solar eclipses as world-threatening disasters.

In ancient Vietnam, solar eclipses were explained as a giant frog eating the sun. The peasantry of ancient Greece believed that an eclipse was the sign of a furious godhead, presenting an omen of wrathful retribution in the form of natural disasters. Other cultures were less speculative in their investigations. The Chinese Song Dynasty scientist Shen Kuo proved the spherical nature of the Earth and heavenly bodies through scientific insight gained by the study of eclipses.

The Eclipse in Native American Mythology

Eclipses have played a significant role in the history of the United States. Before Europeans settled in the Americas, solar eclipses were important astronomical events to Native American cultures. In most native cultures, an eclipse was a particularly bad omen. Both the sun and the moon were regarded as sacred. Viewing an eclipse, or even being outside for the duration of the event, was considered highly taboo by the Navajo culture. During an eclipse, men and women would simply avert their eyes from the sky, acting as though it was not happening.

The Choctaw people had a unique story to explain solar eclipses. Considering the event as the mischievous actions of a black squirrel and its attempt to eat the sun, the Choctaw people would do their best to scare away the cosmic squirrel by making as much noise as possible until the end of the event, at which point cognitive bias would cause them to believe they'd once again averted disaster on an interplanetary scale.

Contemporary American Solar Phenomena

The investigation of solar phenomena in twentieth-century American history had a similarly profound effect on the people of the United States. A total solar eclipse occurring on the sixteenth of June, 1806, engulfed the entire country. It started near modern-day Arizona. It passed across the Midwest, over Ohio, Pennsylvania, New York, Massachusetts, and Connecticut. The 1806 total eclipse was notable for being one of the first publicly advertised solar events. The public was informed beforehand of the astronomical curiosity through a pamphlet written by Andrew Newell entitled *Darkness at Noon, or the Great Solar Eclipse.*

This pamphlet described local circumstances and went into great detail explaining the true nature of the phenomenon, dispelling myth and superstition, and even giving questionable advice on the best methods of viewing the sun during the event. Replete with a short historical record of eclipses through the ages, the *Darkness at Noon* pamphlet is one of the first examples of an attempt to capitalize on the mysterious nature of solar eclipses.

Another notable American solar eclipse occurred on June 8, 1918. Passing over the United States from Washington to Florida, the eclipse was accurately predicted by the U.S. Naval Observatory and heavily documented in the newspapers of the day. Howard Russell Butler, painter and founder of the American Fine Arts Society, painted the eclipse from the U.S. Naval Observatory, immortalizing the event in *The Oregon Eclipse.*

Four more total solar eclipses occurred over the United States in the years 1923, 1925, 1932, and 1954, with another occurring in 1959. The October 2, 1959 solar eclipse began over Boston, Massachusetts. It was a sunrise event that was unviewable from the ground level. Eminent astronomer Jay Pasachoff attributed this event to sparking his interest in the study of astronomy. Studying under Professor Donald Menzel of Williams College, Pasachoff was able to view the event from an airline hired by his professor.

To this day, many myths surround these astronomical events. In India, some local customs require fasting. In eastern Africa, eclipses are seen as a danger to pregnant women and young children. Despite the mystery and legend associated with unique and rare astronomical events, eclipses continue to be awe-inspiring. Even in the modern day, eclipses draw out reverential respect for the inexorable passing of celestial bodies. They are a reminder of the intimate relationship between the denizens of Earth and the universe at large.

Present Day Eclipses

The year 2016 brought the world just two solar eclipses. A total solar eclipse occurred on the 9th of March. An annular solar eclipse, in which the sun appears as a "ring of fire" occurred on the 23rd of March. If you're interested in seeing this rare and exciting solar

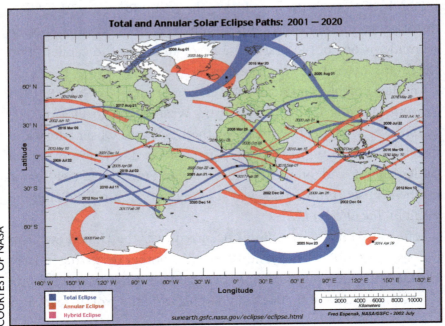

phenomenon yourself, you must to travel to either South America or Western Africa on the 26th of February, 2017.

The next total solar eclipse viewable from the United States, or anywhere else in the world, will occur on the twenty-first of August, 2017. It will be visible in Oregon, Idaho, Wyoming, Nebraska, Kansas, Missouri, Illinois, Kentucky, Tennessee, Georgia, North Carolina, and South Carolina. This event will be the only total solar eclipse for Americans this decade.

Future American Eclipses

The next total eclipse to cross the continental United States is April 8, 2024. It will travel from Texas to Maine. After that, the next American total eclipses will be in 2044 and 2045. No one alive today will see another total eclipse over Grand Teton National Park which will be in 2252.

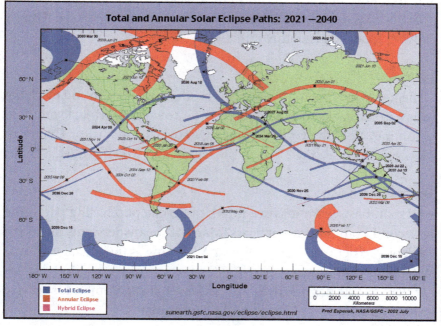

All About Jackson

OVERVIEW OF JACKSON

The town of Jackson is in the northwestern corner of Wyoming. It is situated sixty miles south of Yellowstone National Park and only a few miles south of Grand Teton National Park. Jackson, also referred to as Jackson Hole, was established at the beginning of the 1900s. The town was inaccessible and difficult to reach. Over winter, residents were often snowed in. Visit the Wort Hotel and the Jackson Hole Historical Society and Museum to see what Jackson looked like one hundred years ago.

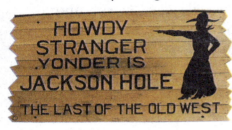

Fast forward to 2017. Access can still be an issue. One year, both the canyon between Hoback Junction and Alpine as well as Teton Pass were closed due to landslides and dangerous snow conditions. The large number of people who commute from Victor, Driggs, and Star Valley were all forced to drive nearly 100 miles one way to work or stay in hotels until the highways were cleared.

Thankfully, during the eclipse, the chance of access being cut off by snow or landslide is highly unlikely.

JACKSON VS. JACKSON HOLE?

The town of Jackson, Wyoming, is located in the south end of a large valley called Jackson Hole. There is no town named Jackson Hole, Wyoming. Visit the Craig Thomas Discovery and Visitor Center in Moose, WY, to see the 3D map the region and why it's called a "hole."

HOTELS AND MOTELS

Jackson has seen a massive influx of people purchasing part-time or second homes in recent years. Many of the owners only stay for a week or two per year, some claiming Wyoming as their taxable "home" state. As a consequence, housing for local workers and lodging for regular tourists has reached the crisis level. It is common for service industry employees to have multiple roommates.

What does this mean for eclipse visitors? Lodging and room rentals in Jackson will be at a massive premium. Does that mean all hope is lost to find a place to stay? Not at all. But you will have to be creative. There will be no (as in zero) hotel rooms available in Jackson by the time this book is printed. The cities and towns around Jackson, and along the path of the eclipse, have been sold out for months.

On December 15, 2016, the author searched on Hotels.com for rooms along the total eclipse path on the weekend of August 21 and found the following:

- Jackson: Snow King Resort for $799 per night.
- Wyoming: No rooms are available in Alpine, Casper, Douglas, Casper, Glenrock, Lander, or Thermopolis.
- Idaho: No rooms are available in Idaho Falls, Pocatello, Blackfoot, Rexburg, Driggs, Victor, or Tetonia.
- Riverton, WY, had rooms available at the Motel 6 for $600 per night. They will be gone by the time you read this.
- Pinedale, seventy-seven miles from Jackson, only had rooms at High Country Suites for $105 per night. Normally the seven other hotels in the town have rooms but are all booked for August 21.

Search for rooms farther away from the eclipse path. If you are willing to stay in Laramie, WY, and drive two and a half hours to Casper, reasonably priced rooms were available in December 2016. As the eclipse approaches, people will book rooms farther from the totality path. By midsummer, rooms in cities like Laramie will feel pressure. The effect of this event will be felt across the country.

Tour buses routinely drive five hours from Salt Lake City to tour Jackson and Grand Teton and then drive back at night. Think regionally when looking for rooms. Be prepared to search far and wide during this major event. If a five-hour drive is manageable, your lodging options greatly expand but it also increases your travel risk.

Yellowstone

Do not stay in towns or cities that require you to drive through Yellowstone to reach Grand Teton or Jackson. Traffic and unexpected animal jams (traffic caused by animals) can delay travel for hours.

Do not believe Google maps (or any other map website) when they show that travel through Yellowstone will only take an hour. Unless you are traveling late at night or early in the morning, add an extra hour to the travel estimate. Bear, wolf, or bison jams are regular occurrences. The 45MPH speed limit in Grand Teton and Yellowstone is strongly enforced.

Internet Rentals

To find rooms to stay in around Jackson, try a web service such as Airbnb.com. Note that many people rent out rooms or homes illegally, against zoning regulations. Property owners make more money with short-term rather than long-term rentals. This fact has led some owners to using their homes as hotels while not complying with safety and rental laws. In response to this, the town of Jackson ramped up enforcement in the summer of 2016. Although zoning laws sound draconian, living next door to a makeshift hotel gets old.

If the town fully enforces zoning laws, authorities may prevent your weekend home rental. Housing in a mountain town popular with outdoor lovers and people who own multiple homes is a challenge.

Jackson is a target for home rental scams. People from out of the area steal photos, descriptions, and addresses, then post the home for rent. You send your check or wire money to a "rental agent" then show up to find you have been scammed. If the deal sounds too good to be true, or it seems strange, trust your instincts.

Camping

If you can book a campsite, do it now. Do not wait. Much of camping in Grand Teton is on a first-come, first-served basis. All areas in the Bridger-Teton National Forest are first-come, first-served. Be aware that the region will be swarming with people. Forest roads will be loaded. Show up early to stake out your spot. Stay farther away in Togwotee Pass, the Wind River Range, Dubois, or farther.

Please respect private land too. Wyoming folks don't take kindly to people overrunning their property without permission. In a big state with only half a million residents, people are very protective, but they're friendly, too. You never know what you might be able to arrange with a smile and a bit of money.

This all said, there are plenty of camping opportunities in north-western Wyoming. You don't have to be exactly in Jackson. If you're ready to rough it, there are national forest camping options.

About twenty government agencies met in the summer of 2016 to talk about how to manage the influx of people. Please note that every possible government agency will be working full time to enforce the various rules and regulations.

NATIONAL PARKS

Chances are you will not find a camping site in the national park. To watch the eclipse from Grand Teton National Park (GTNP), you do not have to sleep in it. You only need to drive there in the morning. Finding campsites in Grand Teton in August is challenging.

Park law enforcement will be present on the eclipse weekend. Parking will overflow. It will make parking lots and lines on Black Friday at the mall look uncrowded. For an event of this magnitude, you'll need to find your parking space near sunrise. Parking will be very limited and roadside parking is not allowed on US 89/191/26. Park in designated areas only. There are entrance fees for Grand Teton and Yellowstone National Parks.

The first sentence of the national parks mission statement is:

> *"The National Park Service preserves unimpaired the natural and cultural resources and values of the national park system for the enjoyment, education, and inspiration of this and future generations."*

Roadside camping (sleeping in your car) is not allowed in the park. Park facilities are only designed to handle so many people per day. Water, trash collection, and toilets can only withstand so much. If you notice trash on the ground, take a moment to throw it away. Protect your national park and help out. Rangers are diligent and hardworking but they can only do so much to manage the expected crowds.

National Forests

There are plenty of forests in the region. They all have camping opportunities. The forest service manages undeveloped and primitive campsites. Be sure to check for any fire restrictions. Check with individual agencies for last-minute information and regulations. The Forest Service requires proper food storage. Plan to purchase food and water before choosing your campsite. Below is a list of national forests within a few hours' drive of Jackson:

Bridger-Teton NF:	www.fs.usda.gov/btnf
Shoshone NF:	www.fs.usda.gov/shoshone
Caribou-Targhee NF:	www.fs.usda.gov/ctnf
Gallatin NF:	www.fs.usda.gov/gallatin

Forest service roads abound in Idaho and Wyoming. Maps are available at the Jackson Hole and Greater Yellowstone Visitor Center at 532 North Cache. Call the Jackson Hole Chamber of Commerce at (307) 733-3316. Visit Jackon's total eclipse site at: tetoneclipse.com.

Printed national forest maps are large and detailed. Viewing digital maps on your smartphone or iPad is difficult. If you plan to camp in the forest, a real paper map is a wise investment.

Camping in federal wilderness areas is also allowed. Those areas afford the ultimate backcountry experience. However, be aware that no mechanized travel is allowed in the specially designated areas. This ban includes: vehicles, bikes, hang gliders, and drones. You can travel only on foot or with pack animals.

Sleep in Your Car

Countless RVs will flood the area. Many local Jacksonites first arrived and slept in the back of their Subarus or trucks. You may find yourself with no other option than to tough it out. Sleeping in your car with friends is tolerable. Doing so with unadventurous spouses or children is another matter.

Do not be caught in the Jackson area without some sort of plan. The whole area brims with people on a normal summer day. August 21 will be anything but a normal day.

Useful Local Webcams

Local webcams are handy to make last-minute travel decisions. The webcams are sensitive enough to show headlights at night. Use them to determine if there are issues before traveling out. It has snowed every month of the year in Jackson and in the surrounding mountains.

The smartphone application Wunderground is useful to check on the webcams in one place. All the webcams are listed near the bottom of the app window.

Weather

It's all about the weather during the eclipse. Nothing else will matter if the sky is cloudy. You can be nearly anywhere in the valley and catch a view of the sky when traffic comes to a standstill. But if there's a cloud cover forecast, seriously reconsider your viewing location.

Travel early wherever you plan to go. Attempting to change locations an hour before the eclipse due to weather will likely cause you to miss the event. Wyoming roads are narrow and slow. The mountain passes are steep, and vehicles routinely back up.

Modern Forecasts

Use a smartphone application to check the up-to-date weather. Wunderground is a good application and has relatively reliable forecasts for the region. The hourly forecast for the same day has been rather accurate for the last two years. The below discussion refers to features found in the Wunderground app. However, any application with detailed weather views will improve your eclipse forecasting skills.

Cloud Cover Forecast

The most useful forecast view is the visible and infrared cloud-coverage map. Avoid downloading this app the night before and trying to learn how to read it. Practice reading them at home. It's imperative to understand how to interpret the maps while still at home.

All cloud cover, night or day, will appear on an infrared map. Warm, low-altitude clouds are shown in white and gray. High-altitude cold

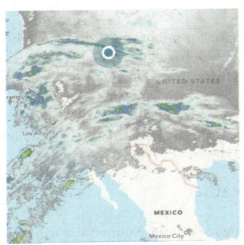

Infrared cloud map showing the worst case eclipse cloud cover.
Courtesy of National Weather Service.

clouds are displayed in shades of green, yellow, red, and purple. Anything other than a clear map spells eclipse-viewing problems.

To improve your weather guess, use the animated viewer of the cloud cover. It will give you a feel for what the clouds are doing. This view gives you an understanding of what is happening. You can discern whether clouds or rain are moving toward, away from, or circulating around Jackson.

Normal Jackson Weather Pattern

Due to the direction of the jet stream, most weather travels across Idaho and Montana, then into Wyoming. On occasion, weather can approach from any direction. Due to the nature of the Rocky Mountains and the Teton Range, weather in Jackson can be unpredictable.

The common weather pattern in August is clear in the morning and cloudy in the afternoon. As the day warms up, clouds develop over Idaho and begin flowing east. These clouds crowd up against the Tetons until the afternoon when they finally flow over. Do not rely on this action though. If anything other than clear skies are predicted, plan to drive to other parts of Wyoming or Idaho to view the eclipse.

Consider that slow-moving clouds can obscure the sun for far longer than the two-minute duration of the totality. The time of totality is so short that you do not want to risk it. Missing it due to a single cloud will be a major disappointment.

Local Eclipse Weather Forecasts

The *Daily* paper will have a weekend edition with articles discussing the eclipse weather. However, conditions change unpredictably in the Rocky Mountains and a three-day forecast may be completely wrong.

The website www.mountainweather.com will feature a paid section specifically focused on eclipse weather during the summer of 2017. Jim Woodmencey runs the site and is a local forecaster. He is regularly featured on local radio stations and at town events.

Forest Fires

For the past several years, forest fires have been common in the western United States. The summer of 2017 is likely to be no different. Forests around Jackson had major forest fires during most of the 2016 summer. Chances are there will be fires in the region in the summer of 2017. For fire updates check https://inciweb.nwcg.gov.

For the best eclipse viewing experience, you need to have as clear a sky as possible. Fog, clouds, or smoke will obscure the subtleties of the sun's corona. If you think the view of the sky is going to be blocked, don't wait until the last minute to move to a clearer location.

Road Closures Due to Fires

Highway 189/191 between Pinedale and Jackson was closed for several days during the summer of 2016. The Cliff Creek Fire, a major burn, blocked the main route into Jackson from eastern Wyoming.

The Lava Mountain Fire blocked highway 26/287 between Dubois and Moran Junction in 2016.

At the end of the 2016 summer, Highway 191 leading from Grand Teton to Yellowstone National Park was closed due to the Berry Fire.

Smoke from the above major events and smaller fires in the region completely blocked the view of the Tetons for over a week. More than once, the noon sun was invisible. An eclipse at that time would have been obscured. Should the above conditions prevail again in late August, Jackson will be a poor place to view the eclipse.

Jackson Hole retains smoke from fires. Often the morning sky is clear, but late-morning winds blow smoke into the valley.

It's imperative to plan for fires and their effects. If strong winds and lightning storms are forecast, prepare to reconsider your viewing location. If conditions are poor, you and thousands of other vehicles will be trapped in slow-moving traffic. Leave Jackson to watch the eclipse at night or extremely early in the morning. RVs are common, and trains of them crawl over Togwotee and Teton passes. Updated road information is available at www.wyoroad.info.

Jackson Information

LOCAL NEWS INFORMATION
People who live in Jackson heavily rely on the *Daily* (as locals call it) paper for updates and news around town. Although people have come to rely on online news sources, the local newspapers are what locals read to find out what is going on. If you want to start a conversation with a local, read the paper and you'll have something to talk about.

THE JACKSON HOLE NEWS AND GUIDE
The *Jackson Hole News and Guide* as well as the *Jackson Hole Daily* paper will have weather and condition updates all week leading up to the event.

The *News and Guide* is a weekly publication that is distributed every Wednesday. It costs $1. The August 16 edition of the weekly paper will be filled with the latest updates. The free daily paper will have the previous weekend's updates, as it's a single paper for Saturday and Sunday.

View the online edition here: www.jhnewsandguide.com.

PLANET JACKSON HOLE
The *Planet Jackson Hole* is a weekly publication produced in Jackson and distributed on Wednesday every week. Billed as "Jackson Hole's alternative voice," it is available throughout the valley: planetjh.com.

TETON VALLEY NEWS
The *Teton Valley Newspaper* will also likely have special sections on weather, with suggestions and condition updates. So if you find

yourself on the Idaho (western) side of the Tetons, the paper is worth looking at. The online edition can be found here:
www.tetonvalleynews.net

BUCKRAIL

Another excellent blog source for updates is the Buckrail News: https://pitchengine.com/buckrail

Google search "Jackson Hole Buckrail News" in case the link doesn't work. This blog has nearly up-to-the-minute information about what is going on in and around Jackson. It often has breaking news.

CELLULAR PHONES

Cellular "cell" phone service in Wyoming is spotty at best. Most of the time there is good coverage along the main highways and interstates. However, even along major thoroughfares, there is little or no coverage. It's possible to find zones where text messages will send when phone calls are impossible. If you cannot make a phone call, the chance of having data coverage for web surfing or e-mail is low.

Please look up any information or communicate what you need before departing from the main roads around Jackson. You may find yourself out of cell service. With a large number of cell users in the valley, coverage and data speeds may collapse. Do a Google search on the phrase "cell phone coverage breathing" to learn more.

JACKSON AREA CELLULAR COVERAGE

As of the writing of this book, Verizon has weak or poor service in the areas south of Jackson past Lockhart Cattle Ranch, through Rafter J and Melody Ranch, and toward the WYDOT. During previous thunderstorms, Verizon has lost cell service. AT&T tends to have better connectivity throughout the valley.

In Grand Teton NP, service may be excellent in some spots and completely dead in others.

- Togwotee Pass has poor to no coverage.
- Teton Pass is generally good.
- Hoback Junction to Alpine, WY, has no coverage.

- The canyon between Hoback Junction and Bondurant is spotty.
- Cache Creek has spotty to no coverage.
- Slide Lake has spotty to no coverage.

Most of the Victor and Driggs areas have good coverage. If you plan to travel into Grand Targhee or any of the mountains and canyons east of Driggs and Victor, you may have little to no connectivity.

Grand Teton National Park Safety

The region around Jackson Hole is full of wild animals. Although they are beautiful, the animals are dangerous. They can easily injure or kill people, as they are far more powerful than humans. Do not feed any wild animals, including squirrels, foxes & chipmunks, as they can carry diseases. These suggestions apply to all public lands in the region. Visit the park's website for more information: www.nps.gov/grte

Bear

Grand Teton National Park and the surrounding forests are bear country. They are home to both grizzly and black bears. Always travel in groups of three or more. If a bear hears you, it will usually vacate the area. Bear charges are frequently caused by surprise.

Noise is the best defense to avoid surprising bears. Regularly clap, make noise, and talk loudly. The *GTNP Bear Safety* brochure states, "Bear bells are not sufficient. The use of portable audio devices is strongly discouraged."

Always have bear spray while hiking. Carry it in a holster or in a cargo pocket. You need to practice being able to discharge the spray within three seconds. Carrying it in a backpack is completely ineffective.

Park regulations require people to stay one hundred yards (300 feet) away from all bears and wolves. They are exciting to see but need their space. Refer to current park regulations for more safety information.

BISON

Bison may appear tame and docile, but they are powerful wild animals. These 1,400-pound herbivores can run three times faster than humans. Bison do not like being crowded or harassed. Stay at least twenty-five yards (75 feet) away to avoid injury or death. Do not walk any closer to capture them with your camera. Do a Google search on "2015 bison gores girl" to see what can happen when people get too close.

MOOSE

This member of the deer family is extremely defensive when they are with their young. If you see a moose calf, leave the area as soon as you can. If a moose approaches, back away. Put something between you and the moose. Unlike bears, it is okay to run from moose. Stay at least 25 yards away from moose at all times.

MOUNTAIN LION

If you encounter a mountain lion, do not run. Keep calm, back away slowly, and maintain eye contact. Do all you can to appear larger. Stand upright, raise your arms, or hoist your jacket. Never bend over or crouch down. If attacked, fight back.

ALTITUDE SICKNESS

Anyone visiting Jackson from a lower elevation might develop symptoms of altitude sickness. The symptoms can run from mild to unpleasant or even dangerous. At Jackson's elevation of 6,200 feet, there is only 81% of the oxygen available at sea level. On top of the Grand Teton, the available oxygen drops to 61%. The lack of oxygen combined with low humidity can cause people to experience altitude sickness.

The symptoms of altitude sickness can strike anyone regardless of fitness level. They include:

- Headaches
- Nausea or vomiting
- Loss of appetite
- Insomnia
- Dizziness
- Fatigue
- Loss of energy

AVOIDING ALTITUDE SICKNESS

1. Hydrate
The number one suggestion to avoid altitude sickness is to drink plenty of water. Consume more than you would at home, especially if you live in a humid environment.

2. Acclimation time
Give your body time to acclimate. Mountaineers who climb high peaks ascend slowly. Give yourself a few days to adjust.

3. Sun exposure
The sun is much more intense in Jackson than at sea level. Wear sunglasses and liberally apply sunscreen to avoid sunburns.

4. Eat well
Keep your energy up. Appetite loss is common at higher altitudes. Maintain your normal eating schedule.

5. Prepare for temperature changes
Temperatures will drop rapidly once the sun sets in Jackson, especially on the mountains. Bring appropriate clothing.

6. Talk with your doctor
If the altitude bothers you or you have experienced previous altitude problems, talk with your doctor before arriving. Seek professional medical attention if you develop serious symptoms.

Eclipse Photography « 27

Viewing and Photographing the Eclipse

AT-HOME PINHOLE METHOD

Use the pinhole method to view the eclipse safely. It costs little but is the safest technique there is. Take a stiff piece of single-layer cardboard and punch a clean pinhole. Let the sun shine through the pinhole onto another piece of cardboard. That's it!

Never look at the sun through the pinhole. Your back should be toward the sun to protect your eyes. To brighten the image, simply move the back piece of cardboard closer to the pinhole. To see it larger, move the back cardboard farther away. Do not make the pinhole larger. It will only distort the crescent sun.

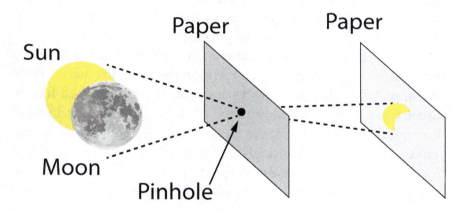

WELDING GOGGLES

Welding goggles with a rating of fourteen or higher are another useful eclipse viewing tool. The goggles can be used to view the solar eclipse directly. Do not use the goggles to look through binoculars or telescopes, as the goggles could potentially shatter due to intense direct heat. Avoid long periods of gazing with the goggles. Look away every so often. Give your eyes a break.

SOLAR FILTERS FOR TELESCOPES

The ONLY safe way to view solar eclipses using telescopes or binoculars is to use solar filters. The filters are coated with metal to diminish the full intensity of the sun. Although the filters can be expensive, it is better to purchase a quality filter rather than an inexpensive one that could shatter or melt from the heat.

The filters attach to the front of the telescope for easy viewing. Remember to give your telescope cooling breaks. Rapid heating can damage your equipment with or without filters attached.

Watch Out for Unsafe Filters

There are several myths surrounding solar filters for eclipse viewing. In order for filters to be safe, they must be specially designed for looking at a solar eclipse. The following are all unsafe for eclipse viewing and can lead to retinal damage: developed colored or chromogenic film, black-and-white negatives such as X-rays, CDs with aluminum, smoked glass, floppy disk covers, black-and-white film with no silver, sunglasses, or polarizing films.

Some online articles state that using developed black-and-white film is safe. Those articles fail to mention the film must have a layer of real metallic silver to protect your eyes. Using developed film is discouraged. You cannot ensure the quality of the film. Feeling no discomfort while looking at the partial solar eclipse does NOT mean your eyes are protected. Retinal damage can occur with zero pain due to the retinas having no pain receptors. Please be careful. Only use protective glasses certified for viewing the eclipse.

Viewing with Binoculars

When viewing the eclipse with binoculars, it is important to use solar filters on both lenses until totality. Only then is it safe to remove the filter. As the sun becomes visible after totality, replace the filters for safe viewing. Protect your pupils. Remember to give your binoculars a cool-down break between viewings. They can overheat rapidly from being pointed directly at the sun even with filters attached.

Planning Ahead

There are many things to keep in mind when viewing a total eclipse. It is important to plan ahead to get the most out of this extraordinary experience.

Understanding Sun Position

All compass bearings in this book are true north. All compasses point to Earth's magnetic north. The difference between these two measurements is called magnetic declination. The magnetic declination for northwest Wyoming in August 2017 is:

11.43°E ± 0.37° for Jackson Hole

Subtract the declination from the azimuth bearing as given in the text, and set your compass to that direction.

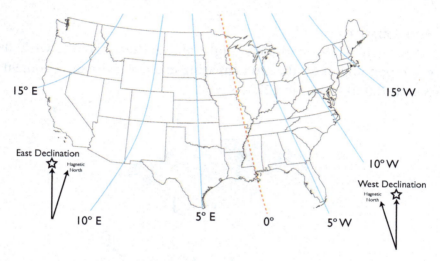

If you purchase a compass with a built-in declination adjustment, you can change the setting once and eliminate the calculations. The Suunto M-3G compass has this correction. A compass with a sighting mirror or wire will help you make a more accurate azimuth sighting.

The Suunto M-3G also has an inclinometer. This allows you to measure the elevation of any object above the horizon. Use this to figure out how high the sun will be above your position.

You can also use a smartphone inclinometer and compass for this purpose. Make sure to calibrate your smartphone's compass before every use, otherwise it might indicate the wrong bearing. Set the smartphone compass for true north to match the book. Understand the compass prior to August 21. There will be little time to guess or search on Google. Smartphone and GPS compasses are "sticky." Their compasses don't swing as freely as a magnetic compass does.

The author has used his magnetic compass for azimuth measurements and a smartphone to measure elevation. Combining these two tools will allow you to make the best sightings possible.

Outdoor sporting goods stores in Jackson and Driggs carry compasses. We recommend purchasing a good compass in your hometown. Take the time to learn how to use it before the day of the eclipse. You do not want to struggle with orienteering basics under pressure.

Sun Azimuth

Azimuth is the compass angle along the horizon, with 0° corresponding to north, and increasing in a clockwise direction. 90° is east, 180° is south, and 270° is west.

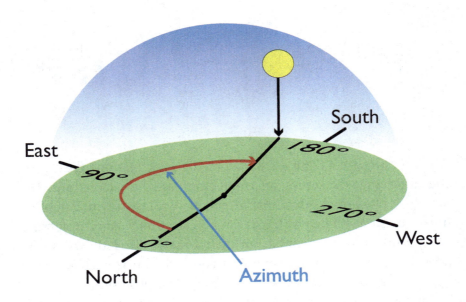

Sun Elevation

Altitude is the sun's angle up from the horizon. A 0° altitude means exactly on the horizon and 90° means "straight up."

Using the sun azimuth and elevation data, you can predict the position of the sun at any given time. Positions given in this book coincide with the time of eclipse totality unless otherwise noted.

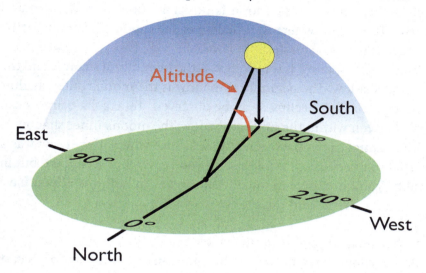

Eclipse Data for Jackson, WY

EVENT	TIME (MDT)	AZIMUTH	ELEVATION
SUNRISE	6:36:00AM	73°	0°
ECLIPSE START	10:16:44AM	113°	38°
TOTALITY START	11:34:56AM	134°	50°
TOTALITY END	11:37:09AM	135°	51°
ECLIPSE END	1:00:28PM	168°	58°
SUNSET	8:17:00PM	287°	0°

Totality Duration: 2 minutes 14 seconds

Eclipse Photography

Photographing an eclipse is an exciting challenge, as the moon's shadow moves at 2,000MPH. There is an element of danger and the pressure of time. Looking at the unfiltered sun through a camera can permanently damage your vision and your camera. If you are unsure, just enjoy the eclipse with specially designed glasses. Keep a solar filter on your lens during the eclipse and remove for the totality!

Partial Vs. Total Solar Eclipse

To successfully and safely photograph a partial and total eclipse, it is important to understand the difference between the two. A solar eclipse occurs when the moon is positioned between the sun and Earth. The region where the shadow of the moon falls upon Earth's surface is where a solar eclipse is visible.

The moon's shadow has two parts—the penumbral shadow and the umbral shadow. The penumbral shadow is the moon's outer shadow where partial solar eclipses can be observed. Total solar eclipses can only be seen within the umbral shadow, the moon's inner shadow.

You cannot say you've seen a total eclipse when all you saw was a partial solar eclipse. It is like saying you've watched a concert, but in reality, you only listened outside the arena. In both cases, you have missed the drama and the action.

Photographing A Partial And Total Solar Eclipse

Aside from the region where the outer shadow of the moon is cast, a partial solar eclipse is also visible before a total solar eclipse within

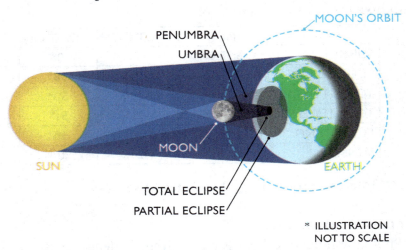

the inner shadow region. In both cases, it is imperative to use a solar filter on the lens for both photography and safety reasons. This is the only difference between taking a partial eclipse and a total eclipse photograph of the sun.

To photograph a total solar eclipse, you must be within the Path of Totality, the surface of the Earth within the moon's umbral shadow.

THE CHALLENGE

A total solar eclipse only lasts for a couple of minutes. It is brief, but the scenario it brings is unforgettable. Seeing the radiant sun slowly being covered by darkness gives the spectator a high level of anticipation and indescribable excitement. Once the moon completely covers the sun's radiance, the corona is finally visible. In the darkness, the sun's corona shines, capturing the crowd's full attention. Watching this phenomenon is a breathtaking experience.

Amidst all the noise, cheering, and excitement, you have less than 150 seconds to take a perfect photograph. The key to this is planning. You need to plan, practice, and perfect what you will do when the big moment arrives because there is no replay. The pressure is enormous. You only have two minutes to capture the totality and the sun's corona using different exposures.

PLAN, PRACTICE, PERFECT

It is important to practice photographing before the actual phenomenon arrives. Test your chosen imaging setup for flaws. Rehearse over and over until your body remembers what you will do from the moment you arrive at your chosen spot to the moment you pack up and leave the area.

You will discover potential problems regarding vibrations and focus that you can address immediately. This minimizes the variables that might affect your photographs at the most critical moment.

It's common for experienced eclipse chasers to lose track of what they plan to do. Write down what you expect to do. Practice it time and again. Play annoying, distracting music while you practice. Try photographing in the worst weather possible. Do anything you can to practice under pressure. Eclipse day is not the time to practice.

Once the sun is completely covered, don't just take photographs. Capture the experience and the image of the total solar eclipse in your mind as well. Set up cameras around you to record not just the total solar eclipse but also the excitement and reaction of the crowd.

ECLIPSE PHOTOGRAPHY GEAR

What do you need to photograph the total eclipse? There are only a few pieces of equipment that you'll need. Preparing to photograph an eclipse successfully takes time. Not only do you have to be skilled and have the right gear, you have to be in the correct place.

BASIC ECLIPSE PHOTOGRAPHY EQUIPMENT

- Solar viewing glasses
- Lens solar filter
- Minimum 300mm lens
- Stable tripod that can be tilted to 60° vertical
- High-resolution DSLR
- Spare batteries for everything
- Secondary camera to photograph people, the horizon, etc.
- Remote cable or wireless release

ADDITIONAL ITEMS

- Video camera
- Video camera tripod
- Quality pair of binoculars
- Solar filters for each binocular lens
- Photo editing software

EQUIPMENT TO PREPARE BEFORE THE BIG DAY

A. Solar viewing glasses

You need a pair of solar viewing glasses as the eclipse approaches.

B. Solar Filter

Partial and total eclipse photography is different from normal photography. Even if only 1% of the sun's surface is visible, it is still approximately 10,000 times brighter than the moon. Before totality, use a solar filter on your lens. Do not look at the sun with your eyes. It can cause irreparable damage to your retinas.

DO NOT leave your camera pointed at the sun without a solar filter attached. The sun will melt the inside of your camera. Think of a magnifying glass used to torch ants and multiply that by one hundred.

C. Lens

To capture the corona's majesty, you need to use a telescope or a telephoto lens. The best focal length, which will give you a large image of the sun's disk, is 400mm and above. You don't want to waste all your efforts by bringing home a small dot where the black disk and majestic corona are supposed to be.

D. Tripod

Bring a stable enough tripod to support your camera properly to avoid unsteady shots and repeated adjustments. Either will ruin your photos. It also needs to be portable in case you need to change locations for a better shot.

E. Camera

You need to remember to set your camera to its highest resolution to capture all the details. Set your camera to:

- 14-bit RAW is ideal, otherwise
- JPG, Fine compression, Maximum resolution

Bracket your exposures. Shoot at various shutter speeds to capture different brightnesses in the corona. Note that stopping your lens all the way down may not result in the sharpest images.

Choose the lowest possible ISO for the best quality while maintaining a high shutter speed to prevent blurred shots. Set your camera to manual. Do not use AUTO ISO. Your camera will be fooled. The night before, test the focus position of your lens using a bright star or the moon.

Constantly double-check your focus. Be paranoid about this. You can deal with a grainy picture. No amount of Photoshop will fix a blurry, out-of-focus picture.

F. Batteries

Remember to bring fresh batteries! Make sure that you have enough power to capture the most important moments. Swap in fresh batteries thirty minutes before totality.

G. Remote release

Use a wired or wireless remote release to fire the camera's shutter. This will reduce the amount of camera vibration.

H. Video Camera

Run a video camera of yourself. Capture all the things you say and do during the totality. You'll be amazed at your reaction.

I. Photo editing software

You will need quality photo editing software to process your eclipse images. Adobe Lightroom and Photoshop are excellent programs to extract the most out of your images. Become well versed in how to use them at least a month before the eclipse.

J. Smartphone applications

The following smartphone applications will aid in your photography planning: Wunderground, Skyview, Photographer's Ephemeris, Sunrise and Sunset Calculator, SunCalc, and Sun Surveyor among others.

CAMERA PHONES

Smartphone cameras are useful for many things but not eclipse photography. An iPhone 6 camera has a 63° horizontal field of view and is 3264 pixels across. If you attempt to photograph the eclipse, the sun will be a measly 30-40 pixels wide depending on the phone. Digital pinch zoom won't help here. If you want *National Geographic* images, you'll need a serious camera and lens, far beyond any smartphone.

Consider instead using a smartphone to run a time-lapse of the entire event. The sun will be minuscule when shot on a smartphone. Think of something else exciting and interesting do to with it. Purchase a Gorilla Pod, inexpensive tripod, or selfie stick and mount the smartphone somewhere unique.

Also, partial and total eclipse light is strange and ethereal. Consider using that light to take unique pictures of things and people. It's rare and you may have something no one else does.

Focal Length & the Size of Sun

The size of the sun in a photo depends on the lens focal length. A 300mm lens is the recommended minimum on a full-frame (FF) DSLR. Lenses up to this size are relatively inexpensive. For more magnification, use an APS-C (crop) size sensor. Cameras with these sensors provide an advantage by capturing a larger sun.

For the same focal length, an APS-C sensor will provide a greater apparent magnification of any object. As a consequence, a shorter, less expensive lens can be used to capture the same size sun.

The below figure shows the size of the sun on a camera sensor at various focal lengths. As can be seen with the 200mm lens, the sun is quite small. On a full-frame camera at 200mm, the sun will be 371 pixels wide on a Nikon D810, a 36-megapixel body. A lower resolution FF camera will result in an even smaller sun.

Printing a 24-inch image shot on a Nikon D810 with a 200mm lens at a standard 300 pixels per inch results in a small sun. On this size paper, the sun will be a miserly 1.25 inches wide!

Photographing the eclipse with a lens shorter than 300mm will leave you with little to work with. Using a 400mm lens and printing a 24-inch print will result in a 2.5-inch-wide sun. For as massive as the sun is, it is a challenge to take a photograph with the sun of any meaningful size.

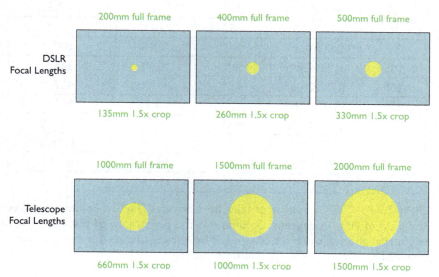

FOCAL LENGTH	FOV FULL FRAME	FF VERT. ANGLE	% OF FF	SUN PIXEL SIZE
14	104° X 81°	81°	0.7%	32.1
20	84° X 62°	62°	0.9%	41.9
28	65° X 46°	46°	1.2%	56.5
35	54° X 38°	38°	1.4%	68.5
50	40° X 27°	27°	2.0%	96.4
105	19° X 13°	13°	4.1%	200.2
200	10° X 7°	7°	7.6%	371.9
400	5° X 3.4°	3.4°	15.6%	765.6
500	4° X 2.7°	2.7°	19.6%	964.2
1000	2° X 1.3°	1.3°	40.8%	2002.5
1500	1.4° X 0.9°	0.9°	58.9%	2892.6
2000	1° X 0.68°	0.68°	77.9%	3828.4

Chart 1: Full-frame camera field of view. The 3rd column is the vertical field of view in degrees. Column 4 is the percentage of the total sensor height that the sun covers. Column 5 is how many pixels wide the sun will be on a 36MP Nikon D810. (Values are estimates)

FOCAL LENGTH	FOV CROP	CROP VERT DEG	% OF CROP	SUN PIXEL SIZE
14	80° X 58°	58°	0.9%	33.9
20	61° X 43°	43°	1.2%	45.8
28	45° X 31°	31°	1.7%	63.5
35	37° X 25°	25°	2.1%	78.7
50	26° X 18°	18°	2.9%	109.3
105	13° X 8°	8°	6.6%	245.9
200	6.7° X 4.5°	4.5°	11.8%	437.2
400	3.4° X 2°	2°	26.5%	983.7
500	2.7° X 1.8	1.8°	29.4%	1093.0
1000	1.3° X 0.9°	0.9°	58.9%	2186.0
1500	0.9° X 0.6°	0.6°	88.3%	3278.9
2000	0.6° X 0.45°	0.5°	117.8%	4371.9

Chart 2: APS-C Crop sensor camera field of view. The 3rd column is the vertical field of view in degrees. Column 4 is the percentage of the total sensor height that the sun covers. Column 5 is how many pixels wide the sun will be on a 12mp Nikon D300s. (Values are estimates)

The big challenge is the cost of the lens. Lenses longer than 300mm are expensive. They also require heavier tripods and specialized tripod heads. The 70-300mm lenses from Nikon, Canon, Tamron, and others are relatively affordable options. It is worth spending time at a local camera shop to try different lenses. Long focal-length lenses are a significant investment, especially for a single event.

To achieve a large eclipse image, you will need a long focal-length lens, ideally at least 400mm. A standard 70-300mm lens set to 300mm will show a small sun. At 500mm, the sun image becomes larger and covers more of the sensor area. The corona will take up a significant portion of the frame. By 1000mm, the corona will exceed the capture area on a full-frame sensor. See the picture on page thirty-seven for sun size simulations for different focal lengths.

SUGGESTED EXPOSURES

To photograph the partial eclipse, the camera must have a solar filter attached. If not, the intense light from the sun may damage (fry) the inside of your camera. This has happened to the author. The exposure depends on the density (darkness) of the solar filter used.

As a starting point, set the camera to ISO 100, f/8, and with the solar filter on, try an exposure of 1/4000. Make adjustments based on the histogram and highlight warning.

Turn on the highlight warning in your camera. This feature is commonly called "blinkies." This warning will help you detect if the image is overexposed or not.

Once the Baily's Beads, prominences, and corona become visible, there will only be 2.25 minutes to take bracketed shots. It will take at least eleven shots to capture the various areas of the sun's corona. The brightness varies considerably. No commercially available camera can capture the incredible dynamic range of the different portions of the delicate corona. This requires taking multiple photographs and digitally combining them afterward.

During totality, try these exposure times at ISO 100 and f/8:

1/4000, 1/2000, 1/1000, 1/250, 1/60, 1/30, 1/15, 1/4, 1/2, 1 sec, and 4 sec.

Photography Time

Set the camera to full-stop adjustments. It will reduce the time spent fiddling. As an example, the author tried the above shot sequence, adjusting the shutter speed as fast as possible.

It took thirty-three seconds to shoot the above 11 shots using 1/3-stop increments. This was without adjusting composition, focus, or anything else but the shutter speed. When the camera was set to full stop increments, it only took twenty-two seconds to step through the same shutter speed sequence.

Assuming the totality lasts less than two minutes, only four shot sequences could be made using 1/3-stop increments. Yet six shot sequences could be made when the camera was set to full stop steps. Zero time was spent looking at the back LCD to analyze highlights and the histogram.

Now add in the bare minimum time to check the highlight warning. It took sixty-three seconds to shoot and check each image using full stops. And that was without changing the composition to allow for sun movement, bumping the tripod, etc. Looking at the LCD ("chimping") consumed **half** of the totality time.

This test was done in the comfort of home under no pressure. In real world conditions, it may be possible to successfully shoot only one sequence. If you plan to capture the entire dynamic range of the totality, you must practice the sequence until you have it down cold. If you normally fumble with your camera, do not underestimate the difficulty, frustration, and stress of total eclipse photography.

Most importantly, trying to shoot this sequence allowed for zero time to simply look at the totality to enjoy the spectacle.

Avoid Last Minute Purchases

You should purchase whatever you think you'll need to photograph the eclipse today. This event will be nothing short of massive. Remember the hot toy of the year? Multiply that frenzy by a thousand. Everyone will want to try to capture their own photo.

Do not wait until the last few weeks before the eclipse to purchase cameras, lenses, filters, tripods, viewing glasses, and associated material. Consider that the totality of the eclipse will streak from coast to coast. Everyone who wants to photograph the eclipse will order at the same time. If you wait until August to buy what you need, it's conceivable that every piece of camera equipment capable of creating a total eclipse photo will be sold out in the United States. Whether this happens or not, do not wait until midsummer to make your purchases. It may be too late.

Practice

You will need to practice with your equipment. Things may go wrong that you don't anticipate. If you've never photographed a partial or total eclipse, taking quality shots is more difficult than you think. Practice shooting the sequence with a midday sun. This will tell you if you have your exposures and timing correct. Figure out what you need well in advance.

Practice photographing the full moon and stars at night. Capture the moon in full daylight. There will be six moon cycles to practice with. Astrophotography is challenging and requires practice.

The May 20, 2012, eclipse as seen in San Diego, CA, shot with a Nikon D300s (crop sensor) with an 80-400mm lens set to ~350mm. The sun is 560 pixels wide on the 4288x2848 image.

This image is shown straight out of the camera without modification. Even with a high-quality camera and lens, photographing an eclipse is challenging. Note the haze and reflection from the overexposed sun.

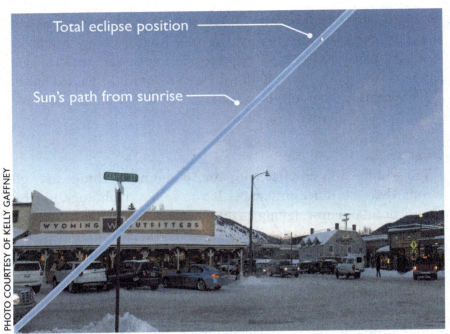

The sun will follow this path on the morning of the eclipse on August 21, 2017. The shot was taken from the corner of Broadway and Center on the town square in front of the elk antler arch.

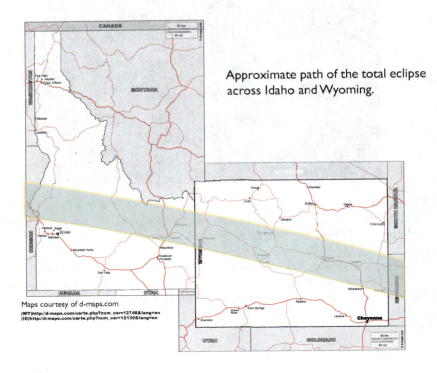

Approximate path of the total eclipse across Idaho and Wyoming.

Viewing Locations Around Jackson Hole

Tens of thousands of people will visit Jackson Hole to view the eclipse. Most will visit the town hoping to see the moon blot out the sun over the Grand Teton from the valley. Due to the sun position when the eclipse happens, this view is impossible. The sun will be in the southeast sky at the moment of totality at 11:35 a.m.

Most of the famous viewpoints in the region will not be in the "correct" direction to photograph the eclipse. If you are looking in the classic directions of the cathedral group, Snake River Overlook, or the Moulton Barn, the eclipse will be behind you, over your left shoulder.

This section contains popular, alternative, and little-known locations to watch the event. As long as there are no clouds or fires, the eclipse will be viewable from virtually anywhere in Jackson Hole.

All distances listed are one way and most listed trails are steep.

SUGGESTED TOTAL ECLIPSE VIEW POINTS

TRAILS
- Cascade Canyon

LAKES
- Slide Lake
- Jenny, String, Leigh
- Taggart, Bradley
- Surprise, Amphitheater
- Lake Solitude

MOUNTAINS
- Snow King
- Glory Mountain
- Jackson Peak

- Sheep Mountain
- Fred's Mountain
- Table Mountain
- Rendezvous Mountain
- Static Peak
- South Teton
- Middle Teton
- Grand Teton

SPECIAL LOCATIONS
- Lamont, ID
- Tetonia, ID
- GTNP Cabins & Districts

Cascade Canyon, South Fork

Elevation: ~8,400 feet, Elevation gain: ~1,600 feet
Distance: 7 miles to the viewing spot one way
Trailhead: Jenny or String Lake parking lots
Difficulty: ★★★★☆

Overview

The South Fork of Cascade Canyon may be the only place to capture the eclipse reasonably close to the Grand Teton. The summit rises 34° above the canyon at the point where the eclipse is directly over the Grand, leaving only 16° between the summit and the sun. However, there is a large ridge between the trail and the summit. Depending on exactly where you are on the trail, you may need to ascend the steep cliff west of the trail to see the Grand.

Missing the Eclipse Risk

MEDIUM. You will need to start hiking early. Clouds flowing from Idaho will not be visible until they flow over the ridge. If clouds move in, you will have little time to find a clear viewing location. Scout this location days in advance. Bear and moose are common.

Photography & Viewing Information

If you can find a point where the summit is visible along the trail, the sun will only be 16° above the peak. You can use an 80-100mm lens on a full-frame camera to capture the Grand and the eclipse. Pro tip: Shoot wider and crop the image later.

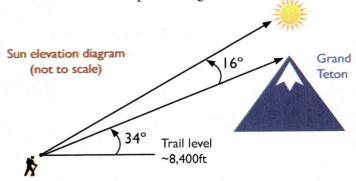

Slide Lake

Elevation:	6,908 feet
Distance:	None, drive up
Trailhead:	Slide Lake parking lot
Difficulty:	☆☆☆☆☆

Overview

Slide Lake was formed by a major landslide in 1925. This lake is located in the Bridger-Teton National Forest. As such, the lake area should not be nearly as crowded compared to Grand Teton National Park. There are no well-maintained trails around the lake.

Missing the Eclipse Risk

LOW. If you decide the location isn't good or clouds come in, you can drive up to twenty-five miles east on the dirt road or return to the highway.

Photography & Viewing Information

You will be able to see the eclipse from any vantage point around the lake. Turn-off spots are plentiful. A photograph with the slide and the eclipse together along the road may be worth investigating.

The information signs along the road to Slide Lake are worth visiting. They document the massive landslide off Sheep Mountain, which created the lake in 1925. The landslide is one of the largest known land movements in history.

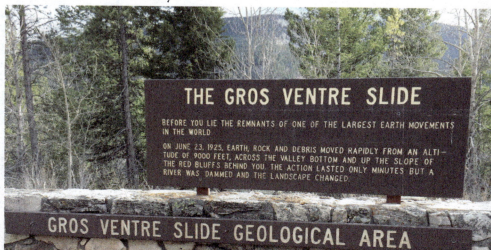

Jenny, String, Leigh Lakes

Elevation: 6,781 feet
Distance: Drive up, or hike up to 2 miles
Trailhead: Jenny Lake or String Lake parking lots
Difficulty: ★☆☆☆☆

Overview

These three lakes are the most accessible in the south section of Grand Teton National Park. In order to see the eclipse from the western side of each lake, you will need to hike from a few hundred yards up to two miles, depending on where you want to view the eclipse.

Missing the Eclipse Risk

LOW. If clouds appear over the Teton Range and threaten to block the eclipse, you may have time to make an adjustment. Consider that thousands of other people will do the same thing. Traffic may come to a full standstill. Make your decision to move as early as possible.

Photography & Viewing Information

Parking will be challenging. South Jenny Lake lot and the buildings around it are under construction. Backpackers, hikers, boaters, and day visitors all use those parking lots.

Expect massive crowds at those locations. If there are clouds, expect a heavy amount of traffic. Everyone will be doing what you are—rushing to view the eclipse at a different location.

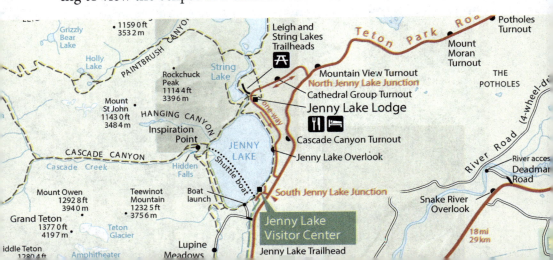

# Taggart & Bradley Lakes

Elevation:	6,921 feet, Elevation gain: 300 feet
Distance:	1.6 miles one way
Trailhead:	Taggart Lake parking lot
Difficulty:	★★☆☆☆

Overview

On low-breeze days, both Taggart and Bradley Lakes will be excellent locations to photograph the eclipse and its reflection. There are no official trails around the western lakeshores. The park encourages visitors to remain on established trails.

Missing the Eclipse Risk

LOW. Should you decide that the lakeshore isn't the right place for you, backtrack toward the parking lot. Except for one stand of lodgepole pine, almost anywhere along the trail is a good spot. If clouds come in, you'll be able to escape. The easier hike to Taggart is along the north section of the lake loop trail. The traffic is higher though.

Photography & Viewing Information

To photograph the eclipse, the western side of either lake may be an ideal location. The hike is not long, so arrive a few hours before the eclipse to scout out a spot.

Surprise & Amphitheater Lakes

Elevation: 9,698 feet, Elevation gain: 2,966 feet
Distance: 4.6 miles one way
Trailhead: Lupine Meadows Trailhead
Difficulty: ★★★☆☆

Overview

Both Surprise and Amphitheater lakes will be memorable spots to watch the eclipse from. The lakes are nestled into the slopes of Disappointment Peak. The views from either lake are unbeatable. The surface of Surprise tends to be mirrorlike due to its protected location. On the southeast edge of Surprise, there is a cliff with nearly vertical views of the valley. The hike to both lakes is steep and sustained. When walking around each lake, take care to avoid stepping on delicate ground cover.

Missing the Eclipse Risk

MEDIUM. Clouds flowing from Idaho will not be visible until they float over Disappointment Peak. Check the weather before venturing out. It takes well over an hour to hike down at a fast pace, so if clouds appear at the start of the eclipse, it may be blocked at totality.

Photography & Viewing Information

The eclipse will be viewable from either lake.

Dangers

Bears are active in the area and are commonly sighted. Signs at Amphitheater warn of this. Have every person in your party carry bear spray.

Lake Solitude

Elevation: 9,032 feet, Elevation gain: 2,982 feet
Distance: 8 miles one way
Trailhead: South Jenny Lake or String Lake parking lots
Difficulty: ★★★★★

Overview

Lake Solitude is the most famous and popular lake in the Tetons that cannot be driven to. The lake is bordered on three sides by massive rock walls rising thousands of feet. Cascade Canyon trail leads away from the lake and up toward Paintbrush Divide. Standing at this lake during the eclipse will put the sun close to the Grand Teton. It's the most easily accessible lake where you can view the eclipse anywhere near the top of the Grand.

Missing the Eclipse Risk

MEDIUM. It will take several hours to reach this lake along a sustained steep trail. Leave early. Do not underestimate the hiking time. Clouds approaching from Idaho will not be visible until they're overhead, leaving you little to no time to run to a better viewing location.

Photography & Viewing Information

From the north shore of Lake Solitude, the total eclipse will be 17° north of and 38° higher than the Grand. Due to the curve of the North Fork of Cascade Canyon, locations closer to the Grand will actually put the sun farther away from the summit. A 24mm lens on a full-frame DSLR should capture the eclipse and summit in one frame.

Pro tip: Try to capture the eclipse, summit, and their reflections in the lake. Test this out days before committing to it.

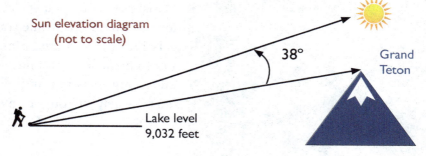

Snow King

Elevation: 7,808 feet, Elevation gain: 1,571 feet
Distance: 2 miles one way
Trailhead: Parking lot at south end of Cache Street
Difficulty: ★☆☆☆

Overview

Snow King is the most accessible mountain in Jackson as it sits at the southern edge of town. It is used by locals for fitness training throughout the year. It is one of the steepest ski slopes in the lower forty-eight states. This is an excellent mountain to acclimate on before venturing to climb other peaks. No camping is allowed on Snow King.

Missing the Eclipse Risk

LOW. Follow the main route to the summit along the heavily trafficked path. You can also ride the chairlift to the top in less than twenty minutes. Navigation is simple, and Jackson is visible for the entire hike. There are other trailheads including Cache Creek Trailhead and Josie's Ridge. Should the weather deteriorate, you may still have time to relocate.

Dangers

Dehydration is the main danger on Snow King. Visitors underestimate the 1,571-foot climb up the mountain. The trail is steep. Trekking poles are recommended.

Most people from low elevations swear the two-mile trail feels like three miles. Make sure to bring at least a liter of water. August is hot and visitors routinely run out of water.

Glory Mountain

Elevation:	10,086 feet, Elevation gain: 1,655 feet
Distance:	~2 miles one way
Trailhead:	Top of Teton Pass, HWY 22
Difficulty:	★★☆☆☆

Overview

Glory Mountain is the easiest peak over 10,000 feet to climb in the Teton Range. The elevation gain is nearly the same as Snow King, but is over 2,000 feet higher. Fit and acclimated climbers can reach the summit in ninety minutes. This high summit provides excellent views of both Idaho and Wyoming.

Missing the Eclipse Risk

LOW. Hike straight up the mountain on the boot pack trail, which follows the southern face all the way to the summit. There is little tree cover, and if you find yourself falling behind schedule, you will have no difficulty watching the eclipse as you climb.

Dangers

Glory Mountain requires no technical climbing. However, the climb is steep and sustained, gaining 1,655 feet in a single mile. If you feel the altitude in Jackson, be ready to breathe hard on this mountain. Pre-acclimate before climbing. Save your knees with trekking poles.

Note: Park at the Teton Pass Summit lot and walk directly north to the trailhead. The parking lot at the top of Teton Pass will be packed with cars. Arrive early in the morning to secure your spot. Another lot is available downhill on the Wyoming side. Use caution walking along the roadway.

Jackson Peak

Elevation: 10,741 feet, Elevation gain: ~2,500 feet
Distance: 6 miles one way
Trailhead: End of Curtis Canyon Road
Difficulty: ★★☆☆☆

Overview

Watch the eclipse from the summit with the same name as the town of Jackson. The hike to this peak in the Gros Ventre Wilderness is enjoyable and easy to navigate. This trail to beautiful Goodwin Lake is marked. Views from the summit are well worth the effort. You can see the Tetons and the Wind River Range.

Missing the Eclipse Risk

MEDIUM-LOW. Hikers can reach this summit in about a six-hour walk. If you plan to catch the beginning of the eclipse on the summit, you will need to begin hiking at no later than four a.m. This peak is heavily traveled. The eclipse track should be visible for the entire climb. If the weather changes, it will be difficult and time-consuming to make a location change.

Dangers

If you continue past Goodwin Lake, the unmarked summit trail requires no technical climbing. Watch for animals. The summit is 5.3 miles from the trailhead and will see heavy traffic.

Note: The road to the trailhead is rough but cuts a lot of climbing out. The peak is popular in the summer, so arrive very early for a parking space. High-clearance vehicle recommended.

Jackson Peak
10,741 feet

Sheep Mountain (Sleeping Indian)

Elevation: 11,239 feet, Elevation gain: 3,696 feet
Distance: 7 miles one way
Trailhead: Near the end of Elk Refuge Road
Difficulty: ★★★☆☆

Overview

The official name of the second most recognizable peak in Jackson is Sheep Mountain, though locals refer to it as Sleeping Indian. It is named for the unmistakable shape of a naturally carved headdress. The actual summit is on the belly of the mountain. The nose is reached from a different direction. An unofficial trail leads all the way to the summit. The eastern face of the summit has a spectacular drop-off. You will be able to watch the moon's shadow sweep across the valley from the top.

Missing the Eclipse Risk

MEDIUM-LOW. The hike to the summit takes roughly six hours. To catch the beginning of the eclipse on the summit, begin hiking at no later than four a.m. The view of the eclipse track is visible for the entire hike. If you are well into the hike and the sky becomes overcast, it will take a long time to make a location change. The summit is deceptively far and there is no water. Don't underestimate this hike.

Dangers

Sleeping Indian requires no technical climbing. Animals are the greatest danger. The summit can be windy. Bring a heavy tripod and additional weights if you plan to photograph.

Note: The road to the trailhead is rough. A high-clearance vehicle is recommended. There is generally parking at one of the trailhead lots during the summer. There is no camping allowed on the mountain.

Sleeping Indian (Sheep Mountain)
11,239 ft

Fred's Mountain

Elevation: 9,862 feet, Elevation gain: 2,011 feet
Distance: 3.2 miles one way
Trailhead: Grand Targhee Resort, use Bannock Trail
Difficulty: ★★★★☆

Overview

Fred's Mountain is accessible from Grand Targhee Resort. Access the resort from Driggs, Idaho. This mountain is unique in that the Teton cathedral group is visible from it. The summit is accessible from the resort. Most uniquely, the eclipse will appear to be almost directly over the Grand Teton. Check with Grand Targhee Resort for access information, as there may be a private event.

Missing the Eclipse Risk

MEDIUM. The hike to the summit is rated "moderate" by the resort. If you are not acclimated, rank this hike as strenuous. If the chairlift is running, ride it to the top. Start hiking by seven a.m. to reach the summit before the eclipse begins.

Photography

Fred's Mountain may be the best location to capture the eclipse above the Grand. The sun will be 17° south of the Grand. Using portrait orientation, a full-frame DSLR with a ~30mm lens will capture the Grand and the eclipse in a single frame.

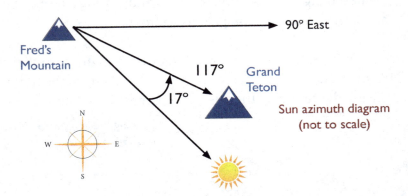

Table Mountain

Elevation: 11,106 feet, Elevation gain: 4,219 feet
Distance: 5.5 miles one way
Trailhead: Teton Campground, reach from Driggs, ID
Difficulty: ★★★★☆

Overview

Table Mountain should be an excellent location to watch the eclipse from. You will have an unobstructed view of the cathedral group and the sun in the same general direction. The final push to the summit is straight up without switchbacks. Bring trekking poles for the climb down.

Missing the Eclipse Risk

MEDIUM. The hike to the summit is long and steep. Start early to ensure you reach the summit before the eclipse begins. The trail is popular with hikers, so do not expect to be alone.

Photography

The sun will be 35° away from the Grand Teton. You will need at least a 24mm lens on a full-frame camera to capture both the Grand and the eclipse. The hike is steep and the summit can be breezy—bringing a tripod will be challenging. Plan accordingly.

Dangers

In 2012, a blogger reported a car break-in at the trailhead. His laptop and tablet were stolen. Apparently, a gang targets cars parked here, as hikers will be gone for hours. Leave no valuables in your vehicle.

Rendezvous Mountain: JHMR Tram

Elevation: 10,450 feet, Elevation gain: 4,139 feet
Distance: 4 miles one way
Trailhead: Teton Village, near the Four Seasons
Difficulty: ★★★★☆

Overview

Riding the Jackson Hole Mountain Resort Tram is the easiest way to watch the eclipse above 10,000 feet. However, all tram tickets sold out in the fall of 2016. The alternative is to hike four miles from the Village to the tram station, watch the eclipse, then ride back down. Strong winds are common on top. Bring a warm coat and gloves.

Missing the Eclipse Risk

MEDIUM. Leave as early as possible. This is a steep climb in bear country. If clouds appear, you may not have time to relocate. However, the eclipse is viewable from the entire trail if you do not reach the summit. Expect hundreds of people.

Dangers

The greatest risks are animals and altitude sickness. You may become stuck if thunderstorms come in. Freezing cold breezes are common on the summit. If you plan to photograph, be prepared for windy conditions. The Village may restrict what you can bring on the tram.

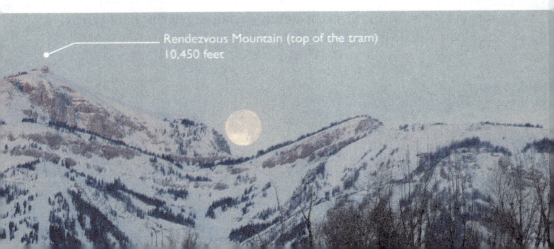

Rendezvous Mountain (top of the tram)
10,450 feet

Static Peak

Elevation: 11,303 feet, Elevation gain: 4,496 feet
Distance: ~9 miles one way
Trailhead: Death Canyon Trailhead
Difficulty: ★★★★☆

Overview

Static Peak is the most accessible of the big peaks in the main Teton group. When looking at the mountain from the valley, the summit is easily recognizable by its sawtooth shape. There is no official trail on the final section of the summit, but it's only a steep off-trail hike. The several-thousand-foot vertical view over the summit edge is spectacular and worth the effort. The hike is steep and sustained.

Missing the Eclipse Risk

MEDIUM. Hikers reach this summit in a single, long day. Starting at three a.m., acclimated hikers will reach the summit around noon. There are no particular navigational challenges while climbing Static Peak. If you find yourself behind schedule and the eclipse starts, find a good clear spot to watch. The biggest risk is getting stuck behind a ridge or in a heavily treed area along the trail.

Dangers

Static Peak requires no technical climbing. Animals are the greatest danger on this trail. The mountain receives a high number of lightning strikes, so watch the weather.

Note: There is generally parking at one of the trailhead lots during the summer. Expect it to fill up on the day of the eclipse.

Static Peak
11,303 ft

South Teton

Elevation: 12,514 feet, Elevation gain: 5,768 feet
Distance: ~7 miles one way
Trailhead: Lupine Meadows Trailhead
Difficulty: ★★★★★

Overview

South Teton is the easiest of the Cathedral Group to climb to see the eclipse. It has the same approach as Middle Teton. The final ascent is a steep hike on the mountain's north face. The dirt is often slick and snow covered, making the climb feel more exciting than one would expect. It's easy to miss the chute on the down climb from the summit.

Missing the Eclipse Risk

MEDIUM. Most climbers can reach the summit in a long single day push. A two a.m. start will put fit, acclimated climbers on the summit around ten a.m. Much time can be lost wandering the boulder field on the approach. Reaching the summit is straightforward. The sun should be visible for most of the climb. If you only reach the saddle between South and Middle Teton late, stop and watch the eclipse.

Dangers

Wear a climbing helmet. This is a high alpine climbing experience with inherent dangers. Rockfall is common. As the morning warms up, the ice melts and releases rocks. Lightning is a danger too. The walk through the boulder field takes a significant amount of time and energy. Do not underestimate this tough climb.

Note: The parking lot is popular and crowded. After summiting and watching the eclipse, consider climbing the Middle Teton if you feel strong and qualified.

South Teton
12,514 feet

Middle Teton

Elevation:	12,804 feet, Elevation gain: 6,058 feet
Distance:	~7 miles one way
Trailhead:	Lupine Meadows Trailhead
Difficulty:	★★★★★

Overview

The Middle Teton can be reached by a scramble. It is an accessible high peak while still providing an enjoyable climb. Once you reach the saddle between Middle and South Teton, plan for at least another ninety minutes of climbing to reach the summit. The Middle Teton has peaks separated by a large gap. It's a great acclimation climb to prepare for the Grand Teton.

Missing the Eclipse Risk

MEDIUM. Most climbers can reach the summit in a single day push. A two a.m. start will put fit, acclimated climbers on the summit around eleven a.m. Reaching the summit before the eclipse will be a challenge. No technical climbing gear is needed for the Southwest Couloir route. The sun should be visible on the entire route. Becoming lost in the boulder field will put you at risk of missing the eclipse.

Dangers

Wear a helmet on this climb, as rockfall is common inside the Southwest Couloir. Lightning is a danger. The walk through the boulder field takes a significant amount of time and energy. Do not underestimate this tough climb. Climbers have died on this mountain.

Note: The parking lot is popular and crowded. Traffic on this mountain is light to moderate. After summiting, spend the extra two hours to bag South Teton.

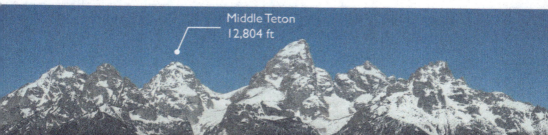

Grand Teton

Elevation:	13,775 feet, Elevation gain: 7,030 feet
Distance:	7 miles one way
Trailhead:	Lupine Meadows Trailhead
Difficulty:	★★★★★★!

Overview

The Grand Teton, tallest of the cathedral group, rises 7,030 feet above the valley floor. Locals refer to this peak as "the Grand." Well over a thousand climbers reach the summit each summer. Exum Mountain Guides and Jackson Hole Mountain Guides are the only permitted guiding services on the Grand.

Missing the Eclipse Risk

HIGH. Most climbers take two days to reach the summit. Backcountry permits are required. A three a.m. start will put fit, acclimated climbers on the summit around one p.m., far too late for the eclipse. The sun **cannot** be seen from the Owens-Spalding route. Any delay will cause you to miss the eclipse. Wear a helmet!

Summer climbing traffic jams are common. If you are behind schedule, descend to the lower saddle to see the eclipse.

Dangers

There are no "safe" routes on the Grand. Wear a helmet. Lightning has killed dozens of climbers over the years. A guide fell and died in the summer of 2016. Do not underestimate the effort and time required. Ice may be present at any time of the year. It is easy to go off-route before reaching the lower saddle. Route finding is a significant challenge. There is 61% of the oxygen available at sea level on the Grand. Do not underestimate this climb.

Note: The parking lot is popular and crowded.

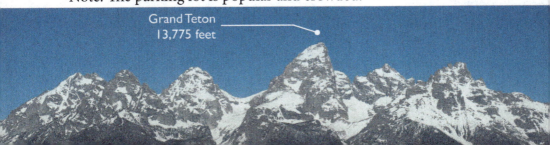
Grand Teton — 13,775 feet

Viewing Locations ☾ 61

Lamont, Idaho

Elevation: 6,030 feet
Trailhead: None, drive up
Difficulty: ☆☆☆☆☆

Overview

Theoretically, there are many locations to view the eclipse directly over the Grand Teton. However, only a few locations are accessible. One of them is in Lamont, Idaho. The Grand is twenty-five miles from this location but it's an easy dirt road drive.

Eclipse start: 10:16:29AM (MDT)
Total eclipse start: 11:34:25AM (MDT)
Total eclipse duration: 1 minute 51 seconds

Missing the Eclipse Risk

LOW. Any clouds that might appear on the western horizon will be easily visible. There are plenty of dirt road and highway escape routes in case of partial cloud cover.

Secret Location

Drive from Tetonia north on the 32. You will drive through a pair of road curves to the town of Lamont. Turn right (north) on N4700E and then turn right (east) on 700N (Coyote Canyon Road). Approximately three miles east, toward the mountains, is an intersection where you can view the eclipse directly over the Grand Teton.

Park along the dirt road for a good vantage point. Driving too close to the mountains will make the cathedral group invisible. You can capture the Grand and the total eclipse in the same frame. The Grand is far away, so it's quite small.

Note: Please respect private property in the area.

Grand Teton, 25 miles from Lamont, Idaho

Tetonia, Idaho

Elevation: 6,047 feet
Trailhead: None, drive up
Difficulty: ☆☆☆☆☆

Overview

Although the total eclipse will not be directly over the Grand, it will hover above the cathedral group from Tetonia. One advantage of viewing the eclipse in Tetonia over Lamont is the mountains are only eighteen miles away.

Eclipse start: 10:16:27AM (MDT)
Total eclipse start: 11:34:19AM (MDT)
Total eclipse duration: 2 minutes 12 seconds

Missing the Eclipse Risk

LOW. Any clouds that might appear on the western horizon will be easily visible. There are plenty of dirt road and highway options in case of partial cloud cover. Expect large crowds.

Secret Location

Tetonia is forty-one miles from Jackson. On a low-traffic day, the drive takes one hour. However, expect the drive to take much longer on eclipse day. Highway 33 is a two-lane road, and traffic is constant throughout the year. Pay attention, as distracted drivers may wander into opposing traffic.

Please respect private property in the area.

Grand Teton, 18 miles from Tetonia, Idaho

Historic Cabins & Districts

Elevation: Various
Trailhead: None, drive up, some walking required
Difficulty: ★☆☆☆☆

Overview

There are several famous cabin locations in Grand Teton National Park to watch the eclipse from. Depending on your photographic skill, you may be able to make an interesting composition with the old buildings. Parking is extremely limited at these sites.

Missing the Eclipse Risk

LOW. All of the cabins are drive-up accessible. Should the sky cloud up, you will still have time to change locations. More information about each of the cabins can be found on the park's website.

Cabins & Barns

- **Luther Taylor Homestead:** Located on Kelly/Gros Ventre Road past Kelly Warm Spring. This property was homesteaded in 1916. The buildings were used in the 1953 movie *Shane*.
- **Moulton Barn & Mormon Row:** The famous barn front faces east, so it's not possible to put the eclipse in a single photo with the barn's front.
- **Cunningham Cabin:** The oldest standing structure in the park, this cabin is located 12.5 miles north of Moose Junction on Highway 89/191/26.
- **Lucas-Fabian Homestead:** Drive into the park, pass Teton Glacier Turnout, and park in the easy-to-miss unpaved parking area on the west side of the road. Walk across Cottonwood Creek.

Please respect these irreplaceable historic structures.

Notes

Alpine, WY, is marginally inside the total eclipse path.

Pinedale, WY, is outside the total eclipse path.

Gannett Peak is the highest point in Wyoming and will experience a two-minute totality. This is a unique opportunity for highpointers.

Your notes: _____

_____

_____

_____

Location	Eclipse Start (AM)	Totality Start (AM)	Totality Duration (min:sec)	Eclipse End (PM)	Totality Azimuth	Totality Elevation	Dist. from Jackson (mi)
WYOMING							
Alpine	10:16:13	11:35:17	0:33	1:00:10	134°	51°	37
Bondurant	10:17:00	11:35:48	1:39	1:01:18	134°	51°	34
Dubois	10:18:11	11:36:50	2:18	1:02:38	136°	51°	86
Gannett Peak	10:17:59	11:36:58	2:02	1:02:51	136°	51°	Hiking
Jackson	10:16:44	11:34:56	2:15	1:00:26	134°	50°	0
Moose	10:16:53	11:34:59	2:19	1:00:29	134°	50°	13
Moran	10:17:16	11:35:33	1:56	1:00:44	135°	50°	30
Pavillion	10:19:15	11:38:25	2:23	1:04:35	137°	52°	151
Pinedale	10:17:31	NONE	NONE	1:02:36	135°	51°	77
Riverton	10:19:33	11:39:03	2:13	1:05:21	137°	52°	163
Wilson	10:16:34	11:34:41	2:15	1:00:12	134°	50°	14
IDAHO							
Driggs	10:16:27	11:34:21	2:18	12:59:42	134°	50°	33
Idaho Falls	10:15:10	11:33:04	1:41	12:58:00	133°	50°	84
Lamont	10:16:29	11:34:25	1:51	12:59:18	134°	50°	54
Rexburg	10:15:42	11:33:14	2:17	12:58:17	133°	49°	81
Tetonia	10:16:27	11:34:19	2:12	12:59:34	134°	50°	41
Victor	10:16:22	11:34:20	2:18	12:59:45	134°	50°	24

☉ is the symbol for the sun and first appeared in Europe during the Renaissance.

☾ is the ancient symbol for the moon.

REMEMBER THE JACKSON HOLE TOTAL ECLIPSE
August 21, 2017

Who was I with? _____
_____
_____
_____
_____

What did I see? _____
_____
_____
_____
_____
_____
_____
_____

What did I feel? _____
_____
_____
_____
_____
_____
_____

What did the people with me think? _____

Where did I stay?_____

Enjoy other Sastrugi Press titles

2017 Total Eclipse State Series by Aaron Linsdau
Sastrugi Press has published several state-specific guides to the 2017 total eclipse crossing over the United States. Check the Sastrugi Press website for the various state eclipse books: www.sastrugipress.com/eclipse/

Antarctic Tears by Aaron Linsdau
What would make someone give up a high-paying career to ski alone across Antarctica? This inspirational true story will make readers both cheer and cry. Eating two sticks of butter a day to survive, the author exposes the harsh realities of the world's largest wilderness. Discover what it takes to pursue a dream.

Lost at Windy Corner by Aaron Linsdau
Climbing Denali is a treacherous affair. Avalanches, blinding blizzards, and crevasses have killed experienced teams. In this dramatic story, Aaron describes the critical choices made during his solo climb and the lessons that were learned as a result. It teaches defining success on your own terms in business and life, with long-lasting messages.

Adventure One by Aaron Linsdau and Terry Williams, M.D.
What does it take to conceptualize, plan, and enjoy your first expedition? This inspirational book contains hard-won knowledge from both authors about their experiences on expeditions around the world. The information provided in this book is useful for any expedition (*Available Summer 2017*)

Cache Creek by Susan Marsh
Five minutes from the hubbub of Jackson's town square, Cache Creek is a popular recreation area on the outskirts of town. It offers the chance to immerse ourselves in wild nature. *Cache Creek* shares some of the ways you can do it.

Visit Sastrugi Press on the web at www.sastrugipress.com to purchase the above titles in bulk. They are also available from your local bookstore or online retailers in print, e-book, or audiobook form.

Thank you for choosing Sastrugi Press.
"Turn the Page Loose"

About Aaron Linsdau

Aaron Linsdau is a polar explorer and motivational speaker. He energizes audiences with life and business lessons that stick. He delivers a message of courage by building grit and maintaining a positive attitude. Aaron teaches audiences how to eat two sticks of butter a day to achieve their goal. He shares how to build resilience to deal with constant pressure and adrenaline overload.

He holds the world record for the longest expedition in days from Hercules Inlet to the South Pole. Aaron is the second only American to complete the trip alone.

This solo expedition is more difficult than climbing Mount Everest with a team. Being alone dramatically increases the challenge. Aaron uses emotionally stirring stories to show how to overcome obstacles, impossible challenges, and unimaginable conditions. He relates these stories to business challenges and shows how the common person can achieve uncommon results.

Aaron collaborates with organizations to deliver the right message for the audience. He relates his experiences to business realities. Aaron loves inspiring audiences. Book Aaron for your next event today.

"NEVER GIVE UP"
GRIT • COURAGE • ATTITUDE • PERSEVERANCE • RESILIENCE

Learn more about Aaron Linsdau at:
www.aaronlinsdau.com or www.ncexped.com.

Smartphone link

Aaron at the South Pole after 82 days alone in Antarctica.

CPSIA information can be obtained
at www.ICGtesting.com
Printed in the USA
FSOW03n0443040417
32683FS